W9-DBX-872

BUILDING BLOCKS OF MATH

NUMBERS

Written by Joseph Midthun

Illustrated by Samuel Hiti

WORLD
BOOK

a Scott Fetzer company
Chicago

World Book, Inc.
180 North LaSalle Street
Suite 900
Chicago, Illinois 60601
USA

For information about other World Book publications, visit our website at **www.worldbook.com** or call **1-800-WORLDBK (967-5325).**
For information about sales to schools and libraries, call 1-800-975-3250 (United States), or 1-800-837-5365 (Canada).

Library of Congress Cataloging-in-Publication Data for this volume has been applied for.

Building Blocks of Math
ISBN: 978-0-7166-4447-7 (set, hc.)

Numbers
ISBN: 978-0-7166-4452-1 (hc.)

Also available as:
ISBN: 978-0-7166-4458-3 (e-book)

Printed in India by Thomson Press (India) Limited, Uttar Pradesh, India
1st printing March 2022

WORLD BOOK STAFF
Executive Committee
President: Geoff Broderick
Vice President, Editorial: Tom Evans
Vice President, Finance: Donald D. Keller
Vice President, Marketing: Jean Lin
Vice President, International Sales: Eddy Kisman
Vice President, Technology: Jason Dole
Vice President, Customer Success: Jade Lewandowski
Director, Human Resources: Bev Ecker

Editorial
Manager, New Content: Jeff De La Rosa
Associate Manager, New Product: Nicholas Kilzer
Sr. Editor: William M. Harrod
Proofreader: Nathalie Strassheim

Graphics and Design
Sr. Visual Communications Designer: Melanie Bender
Sr. Web Designer/Digital Media Developer: Matt Carrington
Coordinator, Design Development and Production: Brenda B. Tropinski
Book Design: Samuel Hiti

Acknowledgments:
Created by Samuel Hiti and Joseph Midthun
Art by Samuel Hiti
Additional art by David Shephard/ The Bright Agency
Additional spot art by Shutterstock
Text by Joseph Midthun

TABLE OF CONTENTS

There is a glossary on page 39. Terms
defined in the glossary are in type **that
looks like this** on their first appearance.

We're the **base ten** system— 10 numbers that make up all other numbers!

You can use us to make any number you can think of!

Yeah, but where do we come from?

Good question!

Humans invented numbers a long time ago.

!

Nope, I'm pretty sure it was humans.

We don't know *exactly* where or when numbers were invented, but we have some good clues...

We *do* know that people didn't always use written numbers.

They likely counted on their fingers when they were out and about.

To keep track of an amount, people etched tally marks on cave walls.

Or a piece of wood!

Or stones!

Or bones!

Each tally mark stood for *one* thing.

Much later, people invented names for different numbers.

Then they started to arrange the names in order by size.

That's counting!

At some point, ancient Egyptians started to use different objects to represent groups of 10 things.

For instance, a single stone might stand for a herd of 10 sheep.

This idea made counting large numbers of things faster and easier!

Today, we continue to represent large numbers by using groups of 10.

That's why we call ourselves the base ten **number system!**

NUMBER WORDS

The numbers 1 through 10 have special names in most languages because people learned to count by using their bodies.

You can count up to 10 on your fingers and then start over.

Let's try!

One. Two. Three. Four. Five. Six. Seven. Eight. Nine. Ten.

Give them a hand!

In English, the words used for the numbers after 10 are based on the first 10 numbers.

ELEVEN!

I'm named after an Old English word, *endleofan*, which means "one left over after ten."

And my name comes from a word that means "two left over!"

After ten!

So, what comes after you guys?

Drum roll, please...

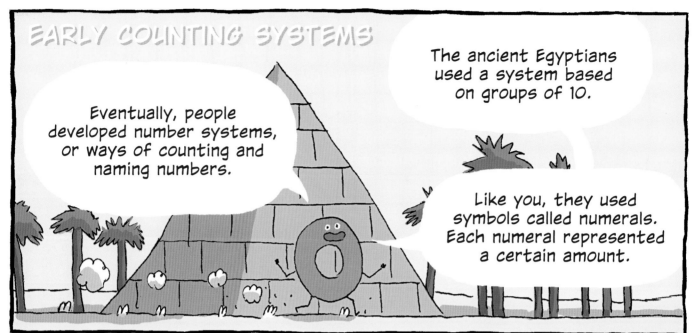

Maybe the Egyptians thought a lotus was the perfect symbol for a big number like 1,000!

Egyptians wrote numerals from left to right, right to left, and sometimes even up and down.

To show more than one of anything...

...I repeat the symbol as many times as needed.

These symbols all mean 23.

crack

CRASH

Twenty-three!

Yikes!

13

Like the Egyptians, the ancient Greeks counted by groups of 10.

But the Greeks used the letters of their alphabet to write numbers.

The first nine letters stood for ones, from 1 through 9.

Α Β Γ Δ Ε F Z Η Θ Ι
1 2 3 4 5 6 7 8 9 10

The ancient Chinese also counted by groups of 10.

They performed calculations using rods made of animal bones or bamboo.

Early Chinese numerals looked like these rods:

1 2 3 4 5
6 7 8 9 10

But base ten isn't the only way people counted.

The Maya of Central America counted using groups of 20.

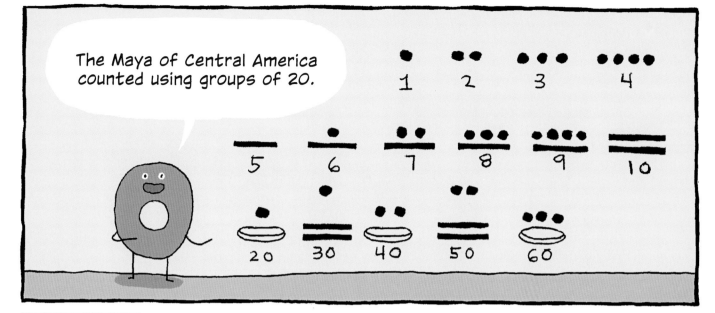

And the **Babylonians** of early **Mesopotamia** counted using groups of 60.

Do you notice anything odd about the symbols for 1 and 60?

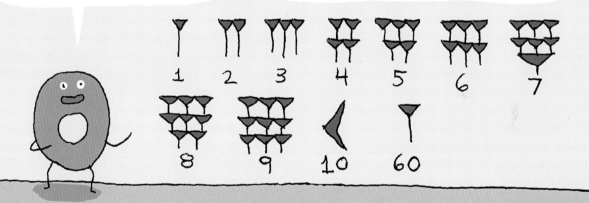

They're exactly the same!

1!

60!

Imagine trying to solve a math problem using the Babylonian system!

ROMAN NUMERALS

Like the Greeks, the **Romans** used letters to represent numbers.

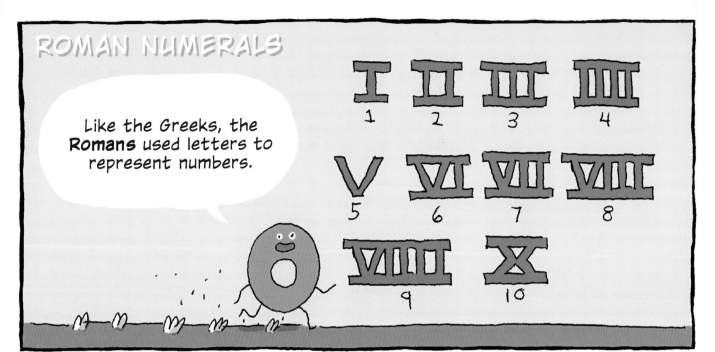

Did you know you can use your hands to make Roman numerals?

The numerals for 1, 2, 3, and 4 look like your fingers.

The numeral for 5 looks like the space between your thumb and finger.

Try it!

And, the numeral for 10 looks like two crossed hands!

After a while, the Romans found a way to save time and space when writing out their numbers.

We used subtraction to make new symbols for the numbers 4 and 9.

The numeral *IIII* became *IV* and the numeral *VIIII* became *IX*...

These new symbols follow a rule:

The smaller numeral goes before the larger numeral to show that it is being subtracted.

The Romans also used other letters to stand for larger numbers...

L = 50
C = 100
D = 500

COUNTING DEVICES

People all over the world developed counting devices to help them work with large numbers.

One of the most popular early counting devices was the **abacus**.

Originally, the abacus was a tray or table covered with dust or sand.

Shake Shake

You could make counting marks with one finger and erase them with a sweep of your hand.

Scritch

Later, people used tables marked with rows or grooves. People used pebbles or beads as counters.

The Romans even had a handheld device—a metal tray with grooves, beads, and a cover for travel.

A simple abacus can still be found in classrooms today.

Each column of beads represents a different group of numbers.

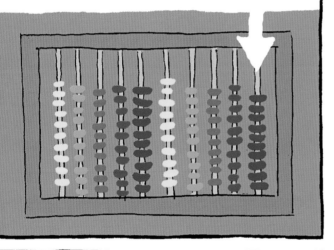

The column on the far right stands for ones.

The next column stands for tens. The third column stands for hundreds.

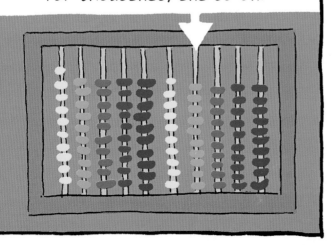

The fourth column stands for thousands, and so on.

So when would you ever need to use an abacus?!

Imagine a distant galaxy in the far future...

On board your spaceship, you have 14 jars of Earth vegetables, *a rare delicacy...*

At a trading post, you decide to purchase 12 more. How many vegetables do you have now?

Normally, you would use your ship's computer to find the answer, but it's not cooperating.

PSH

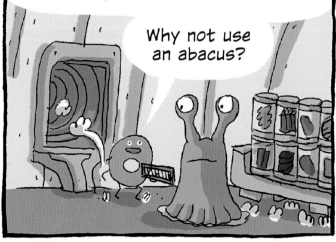

And you're a little bit rusty at doing math in your head!

Why not use an abacus?

To show the 14 jars that you have, move up 4 beads in the ones column and 1 bead in the tens column.

Now, let's add the 12 jars you've decided to buy.

click
click
click
click

Move up 2 more beads in the ones column...

...you've got: 4 + 2 = 6.

click
click

And move up 1 more bead in the tens column. Now you have:

10 + 10 = 20.

Click

Add the 2 tens and the 6 ones together to find out how many vegetables you have in all.

Splrrb! Glp!

That's correct!

26!

That should be plenty of veggies to make a feast!

HINDU-ARABIC NUMERALS

Take a look at these symbols for the numbers 1 through 9...

Look familiar?

In ancient times, the **Hindu** people of India used these numerals.

Over many years, and with some changes, they have become the numerals that you use today!

	1	2	3	4	5	6	7	8	9
A.D. 100-200's	—	=	☰	⅄	ꓱ	ꝥ	𝟕	ら	？
1000's	١	٢	٣	٣	△	५	∨	∧	9
1400's	1	2	3	୪	୨	6	∧	8	9
TODAY	1	2	3	4	5	6	7	8	9

The Arab people learned about these numerals and began using them, too.

After the Arabs conquered most of Spain, they brought the numerals to Europe.

Because of that, the Hindu numerals became known as Arabic numerals.

At the time, people in Europe were using Roman numerals...

...and, they continued to do so for several hundred years.

But that was all about to change.

Across Europe, mathematicians began using Hindu-Arabic numerals instead of Roman numerals.

Hmm, I wonder why...

When you're writing out equations, it's easy to make mistakes.

And using Roman numerals for large numbers can make things confusing.

XXVII + LXIII

squeak

With Arabic numerals, the same equation is easier to use.

27 + 63

That's because the Arabic numerals have **place value!**

PLACE VALUE

So, each *place* in a number has a different value!

An Arabic numeral can mean ones, tens, hundreds, or more, depending on its position in a number.

To better understand place value, let's take a look at some numbers.

In 25, the 5 represents 5 ones.

And I represent 2 groups of 10.

Together, we make 25!

In 50, the 0 stands for 0 ones.

The 5 stands for 5 tens.

Together, we make 50!

In 500, the 0's stand for 0 ones and 0 tens.

And, the 5 represents 5 hundreds.

500!

But Roman numerals, such as V, have the same meaning no matter their position in a number.

Five forever!

After European mathematicians made the Arabic numerals popular, other people in Europe started to use them, too.

SHOWING NOTHING

Wait—I almost forgot to introduce you to one of the most important concepts of mathematics—

Me!

I'm Zero!

You use me to show *no amount.* Zip, zed, nada!

Some number systems had ways to work around using a value for nothing.

The Romans and Egyptians had symbols for 10, 100, and more. But they had nothing for me...

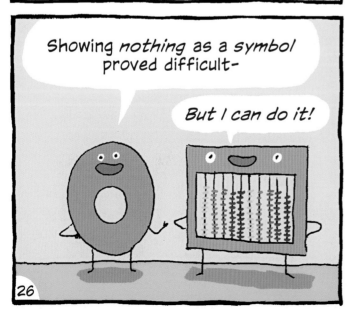

Showing *nothing* as a *symbol* proved difficult—

But I can do it!

For example, to show the number 30, you can separate 3 beads in the tens column...

...and *none* in the ones column.

CLICK

The Maya were among the first people to use symbols for the idea of zero.

Later, the Hindus found a way to show nothing.

They created a numeral and called it sunya.

Sunya stood for an empty column on an abacus.

How enlightening!

By using just 9 numerals and sunya, the Hindus could write *any* number!

The Arabs used sunya along with the other Hindu numerals.

They renamed it sifr, the Arabic word for "empty."

The Europeans adopted the numeral sifr along with the rest of the Arabic numerals.

They renamed it...

ZERO!

TIMELINE

The Egyptians used a hieroglyphic number system based on the number 10.

The ancient Romans created the Roman numeral system.

The Maya of Central America are believed to be the first to use a symbol for the number "0".

3000 B.C.

500 B.C.

250

200 B.C.

2100 B.C.

A Chinese math book had the first known reference to negative numbers.

The Babylonians developed a number system based on the number 60. We still use that system today to count minutes and seconds!

Mathematicians in India developed the decimal system.

595

Italian mathematician Leonardo Fibonacci introduced the Hindu-Arabic numerals to Europeans.

1202

German mathematician Gottfried Leibniz developed the binary number system. Today, most computers are programmed using binary numbers!

1700

628

Indian mathematician Brahmagupta was the first person to describe the number "0" as an actual number (previously, zero had been only used as a placeholder, like in the number 203).

1450

The Hindu-Arabic numeral system came into general use in Europe when the digit symbols were standardized.

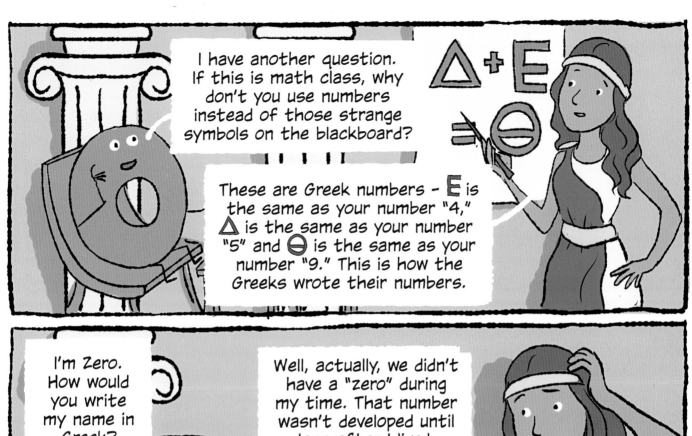

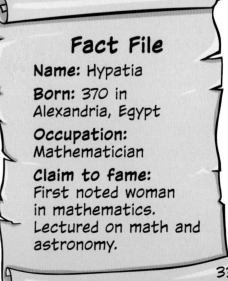

Fact File

Name: Hypatia

Born: 370 in Alexandria, Egypt

Occupation: Mathematician

Claim to fame: First noted woman in mathematics. Lectured on math and astronomy.

ACTIVITY: COUNT LIKE AN ANCIENT EGYPTIAN!

As you saw earlier on pages 12 and 13, the Egyptians used special symbols for writing numbers.

The finger stroke stood for 1.
The arch stood for 10.
The curved rope stood for 100.
The lotus flower stood for 1,000.
The bent finger stood for 10,000.
The tadpole stood for 100,000.

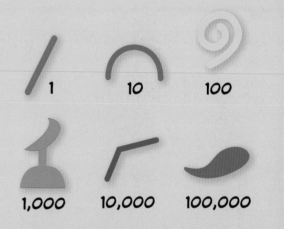

1   10   100

1,000   10,000   100,000

To show more than one of anything, the Egyptians repeated the symbol the correct number of times. For example, to show the number 123, the Egyptians would write 1 curved rope, 2 arches, and 3 finger strokes, like this.

34

Let's see if you can write these numbers in Egyptian! The answers are on page 38.

1. 35
2. 112
3. 1,245
4. 10,437
5. 162,354

ACTIVITY: COUNT LIKE AN ANCIENT ROMAN!

Now that you know how to write Egyptian numbers. Let's see how you do with Roman numerals.

As you saw earlier on pages 16 and 17, the Romans used letters to represent numbers.

I = 1
V = 5
X = 10
L = 50
C = 100
D = 500
M = 1,000

Just like with the Egyptian numeral system, you repeat the letters to show more than one of anything. For example, the number 3 is written as 3 "I's." The number 17 is written as 1 "X," 1 "V," and 2 "I's" (for 10 + 5 + 1 + 1). The number 80 is written as 1 "L" and 3 "X's" (for 50 + 10 + 10 + 10).

III = 3
XVII = 17
LXXX = 80

1,233 = MCCXXXIII

Wow, these numbers can get real long! The number 1,233 is written as 1,000 (the M) plus 200 (the CC) plus 30 (the XXX) plus 3 (the III). Can you imagine how long the bigger numbers would be?

As you also learned, the Romans used subtraction to make the numbers 4 and 9. When a smaller numeral went before a larger numeral, it showed that the smaller number was being subtracted. So, to write the Roman numeral for 4 (IV), you subtract 1 (I) from 5 (V). And, to write the Roman numeral for 9 (IX), you subtract 1 (I) from 10 (X).

IV = 4
IX = 9

So, to write the Roman numeral for 40 (XL), you subtract 10 (X) from 50 (L). And, to write the Roman numeral for 900 (CM), you subtract 100 (C) from 1,000 (M).

XL = 40
CM = 900

OK, now it is your turn. Let's see how well you can use Roman numerals! The answers are on page 38.

1. Write these numbers as Roman numerals:
   a. 26
   b. 84
   c. 351
   d. 1,424
   e. 2,356

2. What numbers do these Roman numerals represent?
   a. XXXIV
   b. XCVII
   c. CCLXXVI
   d. MCCXXXIV
   e. MMMCCCXXXIII

ANSWERS

Count Like an Ancient Egyptian

1. 

2. 

3. 

4. 

5. 

Count Like an Ancient Roman

1. a. 26 = XXVI
   b. 84 = LXXXIV
   c. 351 = CCCLI
   d. 1,424 = MCDXXIV
   e. 2,356 = MMCCCLVI
2. a. XXXIV = 34
   b. XCVII = 97
   c. CCLXXVI = 276
   d. MCCXXXIV = 1,234
   e. MMMCCCXXXIII = 3,333

WORDS TO KNOW

abacus a frame with rows of counters or beads used for adding and other tasks in arithmetic. The abacus was used by the ancient Greeks and Romans and in China and other Asian countries. Today, it is used in schools.

Babylonian having to do with Babylonia, an ancient region in what is now southern Iraq. Babylonia was the site of several kingdoms.

base ten a number system that uses 10 basic symbols: 1, 2, 3, 4, 5, 6, 7, 8, 9, and 0. The value of any of these symbols depends on the place it occupies in the number.

binary number a number written with only two digits: 1 and 0.

Hindu one of a group of people living in India.

Mesopotamia an ancient region in the Middle East where the world's earliest cities were built.

number system a way of writing numbers. People in most parts of the world use the base ten number system.

place value the value of a digit as determined by its place in a number.

Roman a citizen of ancient Rome.

INDEX

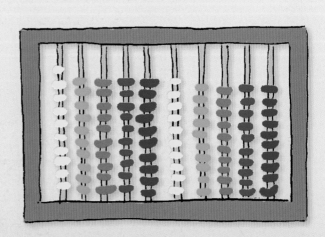